# CHICKENS AT HOME

**by**

***Michael Roberts***

**Illustrated by**
***Sara Roadnight***

Cover photograph
Silver Laced Wyandotte hens by *Richard Roadnight*

Published by Domestic Fowl Research
Printed by Ashley House Printing Company, Exeter
ISBN 0 947870 07 5

# Contents

# INTRODUCTION

Have you ever cooked and eaten an egg which is still warm from being laid? If not, then you have a delicious treat in store discovering the true meaning of 'fresh' eggs from your own hens. Keeping chickens is not only fun but easy when you apply the basic principles of adequate housing, clean drinking water and proper feeding. Don't be put off by horrific stories of chickens catching any and every disease known to science, because common sense and good management are the best disease preventers. You must not forget that chickens require attention seven days a week.

## Non Profit-Making Exercise

Some people think that by keeping a few hens at home they will be on to a good profit-making business. If the birds are kept well, you will show a small profit but mostly the only 'profit' you will have is the pleasure of keeping your own hens and the supply of really fresh, free range eggs.
**NB.** Six hens can produce 3 1/2 dozen eggs per week: make sure you have a ready market for surplus eggs.

## Check Local By-Laws on Keeping Hens

There are local regulations in certain areas preventing you from keeping livestock in your garden. A quick check with your Environmental Health officer at your local Council offices is necessary and don't purchase a cockerel unless your garden is very isolated or there are other cockerels in the area, as cockerels do not lay! A cockerel is not necessary for the production of eggs, and hens tend to be much tamer without one. Also you must think of your neighbours; they are unlikely to appreciate being awakened early in the morning.

## Time Involved

If the system of keeping your chickens is simple and proven, then the amount of time for looking after six birds can be as little as ten minutes per day and perhaps half an hour at weekends. As with all hobbies, however, the more time you can devote to them, the more satisfying they are and the better the results obtained.

**Expenses**

Here is a list of the items you will require for keeping six hens on a Fold system (see method 2). We have taken this system as the quickest and simplest way of keeping chickens.

**Starting up fixed costs**

| | |
|---|---|
| Fold house | £250.00 |
| Galvanised drinker | £22.00 |
| Food hopper | £20.00 |
| Plastic dustbin for food storage | £10.00 |
| Total cost of equipment | £302.00 |

**Variable costs per annum**

| | |
|---|---|
| Six hens @ £6.50 | £39.00 |
| Allow 100g of pellets per bird per day; 6 hens will eat 25kg (1 bag) in six weeks ie. 8$^{1}/_{2}$ bags @ £7.50 | £63.75 |
| Wheat 100kg | £26.00 |
| Grit, bales | £7.00 |
| Total cost of birds' food, etc. | £135.75 |

| | | |
|---|---|---|
| Layers pellets | £7.50 | per 25kg |
| Wheat | £6.50 | per 25kg |
| Straw (wheat) | £1.00 | per bale |
| Mixed grit | 50p | per kg |
| Shavings | £6.00 | per large bale |

**Benefits from six hens per annum**

Maximum production from a hybrid hen is 280 eggs per annum

Sale of surplus eggs 1680 ÷ 12 = 140 doz. @ £1.50 = £210.00

6 hens will produce about 8 x 25kg bags of manure per annum @ £3.50 per bag

+ 6 hens @ £1.50 (at end of lay) = £9.00

+ garden manure = £28.00

£247.00

Assuming the family will eat a dozen eggs per week = 52 doz @ £1.50 = £78.00

\- £78.00

£169.00

# WHICH TYPE OF HEN DO I BUY?

For most beginners it is far better to start with a commercial breed such as Comet, Warren, Black Rock, etc. rather than a pure breed as

a) they cost less, therefore if you lose one it does not hurt the pocket quite so much.
b) they are usually quieter in temperament, and
c) they lay better than most pure breeds, but not for as many years and of course they are not so attractive.

The cost of a commercial breed or hybrid at point-of-lay (POL), normally somewhere between 18-22 weeks old, is £5.00 - £7.00 per bird.

# WHERE DO I GET MY CHICKENS?

**Commercial or hybrid hens:** first check your local newspaper, or buy a copy of **Poultry World** (you may have to order this from your newsagent) or **Farmers Guardian**. Cyril Bason from Craven Arms, Shropshire, will supply small numbers of birds and will deliver them. Make sure they have been reared outside.

# PURE BREEDS

If you decide to go for a pure breed, then you have about 160 to choose from. The problem is finding the breed you want and the numbers you require. The most easily available pure breeds are **Rhode Island Red** and **Light Susssex**. Both are good layers, then there are brown egg layers like **Marans, Welsummers** and **Barnevelders**, all these birds originating from Holland. There are white egg layers like **Leghorns, Anconas, White Faced Spanish, Andalusians, Minorcas, Sicilian Buttercups** - all Mediterranean breeds as their names suggest. White eggs are more popular in America than England. Also, all hens with white ear lobes lay white eggs. All these breeds are prolific layers and the **Minorca** lays a huge egg. These breeds are only suitable for free range systems. **Derbyshire Redcap** and **Old English Pheasant Fowl** are two good British egg layers, again, free range breeds. **Buff Orpington** (also in White, Black and Blue (grey)) are worth considering. In Canada there are still huge flocks of commercial White Orpingtons. **Wyandottes** - the range of colours is fantastic: the normal egg layers are White but the Silver or Gold Laced are very attractive.

There are many continental breeds which are good layers such as **Silver Campines** and **Appenzeller Spitzhauben**, all with their various merits which are stoutly defended by their breeders.

For colour pictures of most of the more common breeds see *British Large Fowl* and *Bantams in Colour* from Domestic Fowl Research. Most large fowl have minitures so the only difference is in size (see back cover).

## WHERE DO I FIND THESE PURE BREEDS?

There are hundreds of small poultry keepers or fanciers dotted around the country with all sorts of breeds, including bantams. The difficulty lies in trying to track them down! Most of these people shed their surplus stock in the late summer and autumn and they can be found in poultry tents at Agricultural shows, Poultry Shows or the Rare Breeds Survival Trust Show and Sale; another option is to join the Poultry Club.

There are several large breeders of pure breeds around the country including The Domestic Fowl Trust, and these can be located through the **Poultry World, Farmers Weekly, Farmers Guardian** and **Fancy Fowl.**

Prices can vary from breeder to breeder; don't go for top exhibition stock as they look fantastic but most of them are likely to be poor layers - the breeders have concentrated on the feathers at the expense of the original purpose of the bird.

It is best to buy your stock at point-of-lay (18-22 weeks old). They will not have laid an egg so do not expect them to start laying as soon as they arrive. Depending on their age, they will need from a few days to a few weeks to settle into their new home and become accustomed to their new diet.

It is often difficult to buy more than a trio of the above birds; the cock bird is nearly always related but there are specialist breeders about who will sell you what you want. It is best to go somewhere else for a cock bird if you want to breed from them. One word of advice: never buy blind, ie. always go and collect your birds to see what conditions they are kept in; do not be afraid of turning away if the stock looks unhappy and unkempt, and never buy from dealers who advertise all sorts of breeds and send them unseen.

Cost of pure breeds can vary from £15 - £40 per bird.

# METHOD 1

## **Minimum Space - Verandah System** (fixed)

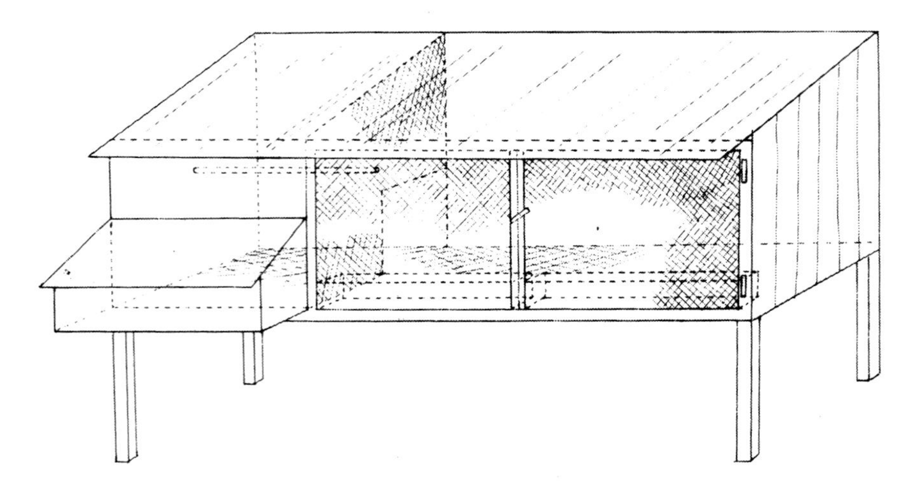

*Verandah house suitable for three birds, in a fixed position*

The Verandah system works well for those people who have a very small garden or yard and who would not ordinarily keep hens due to lack of space. The house described is only one of many designs which can be constructed by the amateur carpenter. The trick is to leave sufficient room for the birds to roost at night, ie. 9 inches per bird, and sufficient space for all the birds to eat at the same time.

Always place the house in a draught-free corner of your garden. Hens can withstand very cold weather as long as they are dry and out of the wind.

Note that the droppings can either be allowed to fall onto the ground or you can place a droppings tray under the house so that you can dispose of the droppings either in your own garden or into the dustbin.

Note all the sections of the house, the covered area with door which contains the perch and nest box, and the open area which contains the food troughs and water drinker. Ideally, the covered area should be one third of the open area with a pop-hole in between.

### Management of birds in a Verandah System

Because your birds are in a confined space, looking after them is most important. Some of the best exhibitors' birds live under such conditions, but the birds can become bored through lack of proper feeding, greenstuff and cleanliness.

**Open Area**

**Feeding:** dry layers pellets in a hopper or trough are best. Layers pellets are vital for good egg production. Every week add some titbits, such as wheat, maize or sunflower seeds, into the food hopper or trough . Clean the hopper out thoroughly once a week. Mash is not only wasteful as it does not look like food once on the ground, but it rapidly fouls the water as it sticks to the beak.

**Water:** there should be a separate trough for water. This must be filled with clean tap water every day, particularly in frosty weather, and the trough should also be rinsed out every day, prior to refilling with clean water. Water troughs become very unhygienic if left for days on end. A bird might mess in it by mistake or the residue from its beak after feeding can start a bacterial infection. The sun causes green algae to grow which is not beneficial, so be sure that the water is out of the direct rays of the sun.

**Feeding greenstuffs:** as your birds under this system have no access to grass, it is important to feed them 'greenery'. This can be lettuce leaves, grass, cabbage, cauliflower outer leaves, apples, swedes. Supply an extra trough for feeding this greenstuff to them, don't give them too much at once, and clear up the trough and pen after two days. You will find that your birds will have a liking for something in particular, like cold baked potato cut in half, or suet, but try to vary their diet as much as possible. Above all keep them busy; a busy hen will not develop any vices due to boredom, such as feather-pecking or egg- eating.

**Grit:** a small grit hopper must be supplied. This is essential for the birds' good health, digestion and egg-laying. Make sure this is in a dry place and clean it out once a week.

**Covered area**

**Nest Box:** collect eggs daily and if you are interested, make a small calendar to record the number of eggs laid each day. Ensure that the nest box is always filled with clean straw or sawdust and once a week empty the whole nest box and put in some clean straw or sawdust, together with some louse/flea powder sprinkled over the top. If there is a broken egg, remove it as quickly as possible and replace the soiled litter.

**Daily Routine**

1. Check the birds, making sure that they look quite happy.
2. Check the food hopper, top up if necessary.
3. Take the water trough out, clean and replace with fresh water.
4. Collect eggs.
5. Add any titbits and/or greenstuff to the food hopper.
6. Ensure that the bottom of the Verandah is clear of any debris from greenstuff from the day before.

**Weekends**

1. Clean droppings tray and replace.
2. Clean nest box; put old litter in the dustbin or on the compost heap. Don't forget the flea powder in the fresh litter.
3. Clean the grit hopper and top up with fresh grit.

Once a year, and especially before the arrival of any new birds, clean and disinfect the house. Remove the birds and dismantle as much of the house as possible. Make sure that all marks and droppings are removed and scrub the entire house, inside and out, with Antec Longlife 250S disinfectant. The house now looks as good as the first day you used it and all the harmful bacteria have been killed. Make sure the house dries thoroughly. Once a year paint all surfaces with Cuprinol to preserve the material. Creosote is highly toxic to birds so it is safer to use Cuprinol or Timber Care or one of the proprietory brands of specified non-toxic wood preservatives.

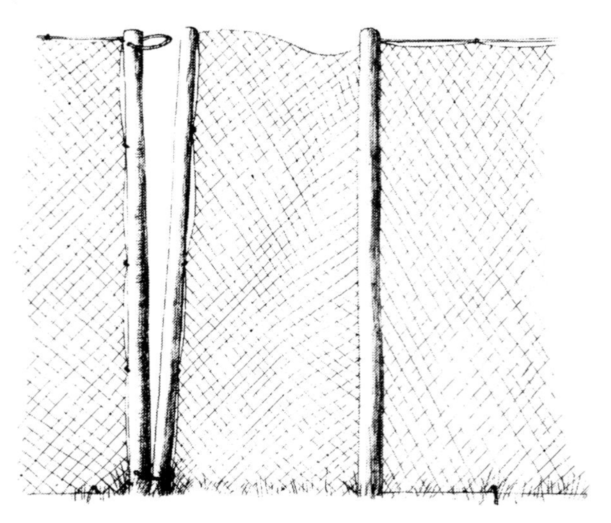

*A simple gate in poultry fencing*

# METHOD 2

**Small garden - Fold System** (movable)

*Fold Unit suitable for 8-10 birds, movable on your lawn. Minimum size of lawn or grassy area required is 20ft x 40ft, preferably flat, and the Fold Unit moved every day: in 3 weeks you will be back where you started.*

This way of keeping hens in a small garden is most satisfactory. The birds always have access to grass, and yet they are contained, away from the garden, the dogs, cats and foxes.

By regular moving of the Fold, depending on the number of birds you have in the house and the time of year, you can still keep a good lawn without ugly muddy or yellow patches and you will be manuring the grass at the same time.

A Fold house consists of a covered area which is approximately one third laying part and two thirds roosting part on a slatted floor, a run section with access to water and a small covered area at the end of the run section for the food hopper. The Fold unit is normally 10ft long and 4ft wide - much larger and it becomes rather more awkward to move. One note here: hens still like perches to roost on, even with a slatted floor, through which all their manure drops. Perches (9" space per bird) should be 2" x 2" with rounded top edges. The slats should be 3/4" x 3/4" and 1 1/4" spacings between. In order to make this unit entirely fox proof, chicken wire netting or 2" wire mesh can be stapled to the floor. The hens will have room to peck and scratch through this wire.

**Management of hens in Fold system**

It is important that your hens should receive regular attention precisely because they are contained. Hens will dehydrate very quickly and die in less than 48 hours if they have no water.

**Feeding:** apart from some natural food they will pick up from the grass, a food hopper should be attached to the wall of the fold in the covered area of the run section. This should be filled with layers pellets and wheat. From time to time, take this home and wash it thoroughly as the feed when damp tends to cake to the sides of the hopper and go mouldy. Mouldy food is a breeding ground for harmful bacteria. Pellets are better than mash as they still look like food when they are on the ground. Mash sticks to the beak and rapidly fouls the water.

**Water:** in order to be able to move the Fold with ease, a metal galvanised drinker (1 gallon for 4-6 birds) suspended from the top of the unit inside and about 4" off the ground is ideal. Clean fresh water is essential every day. Empty at night if there is likely to be a frost; this will save time in the morning. Hens do not feed or drink in the dark.

**Grit:** essential for the bird's good health, digestion and egg laying: a hopper should be placed about hen shoulder height near to the food hopper in the covered area of the run section.

**Feeding greenstuffs:** this is not necessary as your birds will find enough extras without additional greenstuffs if they are moved regularly. There is nothing to stop you giving your birds extra greenery, but it can make the Fold run section look very untidy. Hang any greenstuffs from the roof of the run.

**Daily Routine**

1. Check your birds, make sure they look quite happy.
2. Move the fold. The best way to do this is to keep the birds locked in their slatted covered area and release them when you have made the move; this way toes and legs do not become pinched or broken.
3. check the food hopper and top up if necessary.
4. Unhook the water drinkers, clean and replace with fresh water.
5. Collect the eggs.

**Weekend Routine**

1. Clean out nest box and put in fresh litter, straw or sawdust, not forgetting a sprinkle of flea powder.
2. Clean out slatted area if necessary. If there is a build up of manure between the slats or in one corner, remove this.
3. Check the grit hopper, clean and top up with fresh grit.

This fold unit will want a thorough cleansing once a year with Antec 250S and a good painting with Cuprinol once a year if the wood has not already been pressure treated. Avoid creosote as it is toxic to birds. You will have to remove the birds while this is going on, and try and choose a warm day to get maximum penetration from the wood preserver.

**Daily Routine**

1. Open the pop hole and let the birds out.
2. Check your birds, making sure they look happy and are all out of the house. If not, check to see why not; they could be laying or moping, unwell.
3. Check the food hopper and top up if necessary. Make sure the feed is flowing evenly all round the hopper.
4. Remove the water drinker, clean and replace with fresh cold water.
5. Collect the eggs, making sure there are no broken ones; if there are, remove nesting box material and replace with clean.

**Weekend Routine**

1. Clean out nest box and put in fresh litter, straw or shavings, not forgetting a sprinkling of flea powder.
2. Check the grit hopper, clean and top up with fresh grit.
3. Turn litter in house, or clean out every two weeks. This will depend on how many birds there are in the house, and cleaning might be stretched to a month, if there are only a few birds in a large house.
4. From time to time it is a good idea to dig lightly over the winter run, to fill the dustboxes and even out the ground. Always ensure that the run is clear of any rotten cabbage leaves, etc. and keep as tidy as possible. Hens are very untidy birds.

The house will want a thorough clean out every year. Strip everything out of the house, perches and all, disinfect with Antec 250S, Cuprinol inside and out. During this operation lock the birds into the summer run with food and water.

Every year, too, Cuprinol the fencing posts, and in order to preserve the wire it is possible with the aid of a paint roller, to coat the wire netting with a black or green bituminous paint. This will double the life of your perimeter fence, and should be repeated every four years. (See page 29).

# METHOD 4

**Deep Litter System**
Many people who buy older country properties find themselves with outbuildings which can be used for keeping chickens in. One advantage of keeping birds in a stable or cowshed is that they are safe from the fox, both four-legged and two-legged. How many can you keep? Allow one bird per two square feet. Remember also the egg factor: don't keep too many hens if you have no market for the eggs. This system has the added advantage that there is no weekly cleaning, only one good clear out per annum.

The birds must have sufficient daylight to be able to feed and drink. The biggest advantage with outbuildings is that they normally have electricity already installed. This is very important especially in winter months when daylight hours can be increased by an electric bulb on a time switch (see General Management). Assuming that there is enough light, the next most important consideration is ventilation. There are several ways to achieve this: knock some bricks or stones out of the wall about a foot from the roof; saw three inches off the top of the door; make a wooden frame with 1/2" wire mesh to fit the window frame and take out the top panes of glass, or open the window and cover the opening with a frame. Ensure that there are no draughts and keep the ventilation as high up as possible. All ventilation must be adjustable for windy days or hot summers, etc.

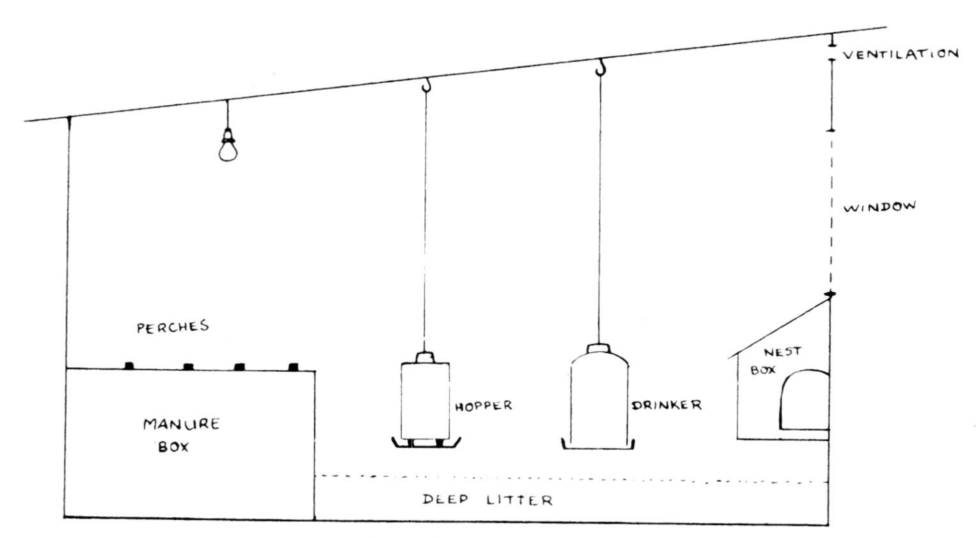


*Deep litter house layout*

The way you lay out this building will affect the production of eggs so the following points are very important:

1. **Nestboxes:** always place them in the darkest corner - the best place is under the window. Please see diagram for nestbox design and measurements and allow one nestbox space per four hens. Aim to give the hens as much privacy and darkness as possible - a hen likes to hide her eggs, so by making a simple, dark place for her to lay her eggs you can get her to lay in a place which is convenient for you when collecting them.

*Allow 12" x 12" for each nesting space*

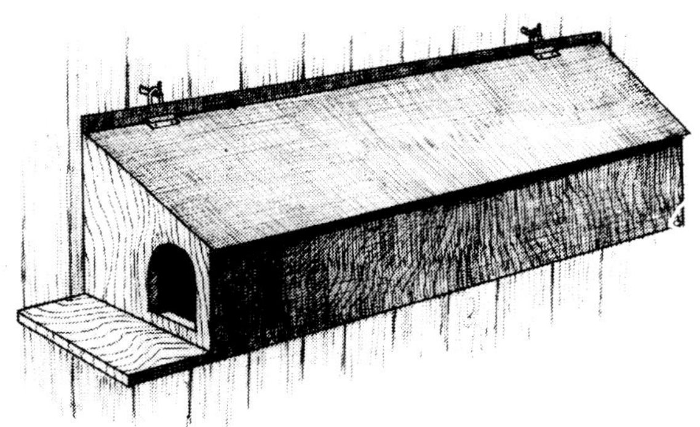

*This model has no partitions for easier cleaning - note entrance at side for maximum darkness.*

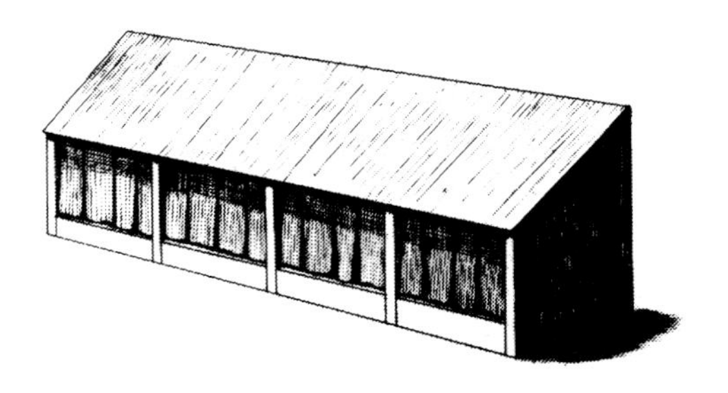

*Nestboxes with black polythene strips for privacy*

2. **Perches:** allow 7" to 12" per bird along the perch, depending on the breed, eg. Leghorns or Comets need 7", Black Rock 9", Cochins 12" and 5" for Bantams. It is always better to have too many perches than too few, and all perches must be the same height, roughly 2ft - 3ft above the ground. For a small number of birds a simple length of 2" x 2" timber, planed and top edges rounded, will be sufficient. Construct a wooden ledge about 6" below the perch to collect the droppings. This ledge can be cleaned once a week and a little sawdust on it makes cleaning quicker and easier.

For a larger quantity of birds, construct a frame or box with corrugated iron sheets around the sides and 2" x 2" perches on the top. Two thirds of the manure produced by a hen is at night so tack some chicken wire (2" mesh) under the perches so that she doesn't fall in to a sewerage box. Another tip is to place some polythene sheeting in the base of the box so cleaning is easier. The perches should face the window and be 18" apart.

3. **Food:** this is best given in the form of layers pellets in tube feeders. These can be hung from the roof or placed on a concrete block or bricks. Always ensure that the level of the feeder trough is the height of the back of your hens - too low and the hopper becomes filled with scratchings, straw, etc., too high and the hens have difficulty in feeding. The hoppers should be placed away from the perches and nestboxes and in plenty of light. Check that there are enough hoppers for all the birds to feed at one time: approximately 20 per tube feeder or 4" trough space per bird. As with drinkers, it is best to use proper equipment rather than make do with bowls, buckets or other household utensils - hens can be very wasteful and tube feeders are designed to minimise wastage of feed. Mash is wasteful as not only does it not look like food once on the ground, it sticks to the beak and rapidly fouls the water. Whole wheat can be fed half and half with the pellets or scattered on the litter to make the hens scratch.

4. **Water:** there are several ways of supplying water to your birds depending on the numbers. For small quantities, a simple metal water font or drinker suspended from the roof or placed on a few bricks (for the same reasons as the food hopper) will do, a 2 gallon drinker for every 25 birds. Or if you are going for a larger quantity of birds, a low pressure water drinking system saves a lot of time water carting. There are several makes on the market: Big Dutchman, Eltex and Lubing and they all work on the same system of a small header tank with a stop cock from the main supply. From the header tank (approx. 20 gallons) a small diameter plastic tube feeds a series of red plastic drinkers that automatically top themselves up as the water is consumed; again, position about hen back height.

They will need regular checking to ensure they are working correctly.

5. **Litter:** what to use on the floor? There are various materials and what you choose depends on what you can obtain most easily or cheaply:

a) wood shavings: these can be bought from the local joinery workshop or timber merchant, also obtainable especially packed in large plastic bags. Be sure to obtain only untreated shavings.

b) wheat straw: about 75p to £1 per bale. ALWAYS remove the string otherwise you will find a hen with its feet tangled and bound.

c) peat moss and sawdust: peat moss from your local garden centre and sawdust from your local timber suppliers. Use about half and half, can be rather dusty.

d) bark pulp and shredded paper can also be used but they are not easily found. NEVER use hay due to harmful mould arganisms present.

The litter should start at about 5" deep and build to about 12". You will find that with regular raking and turning over, the litter makes its own kind of compost, breaking down the manure and staying friable, clean looking and odour free.

6. **Grit:** with a small quantity of birds a small grit hopper attached to the wall or door is a must. With a larger flock, sprinkle the grit by hand liberally around the litter once a week; about a tea mug of grit per 25 birds.

At the end of the laying season, normally August - September, a thorough clean out of your building is necessary. This means the removal of all perches, perch fittings and droppings (sewerage) box, nestboxes, food hoppers, drinkers and grit hoppers, if used. Clean out the litter and this can go onto the vegetable garden (peas and beans especially) or compost heap. Then wash all the walls, floor and ceiling with Antec Longlife 250S, scrubbing off all the caked manure - the building must look like new, and be allowed to stand empty for a week or two. Clean all hoppers, drinkers, the droppings box and shelf, with the same disinfectant. The perches and nestboxes should be painted with Cuprinol plus all wooden parts in the outbuilding. Hens do suffer from parasites, mainly red mites and lice. Wood offers an ideal hiding place for these irritants, especially wood with cracks or knot-holes, or bark on it (remove bark before using). Spray with Duramitex. (See page 23).

**Greenstuff**

Most commercial deep litter chicken farmers don't bother with greenstuff, but greenery makes the eggs taste and look better and also gives the hens something to do. A cabbage or sprout plant hung up by a length of cord off the ground is ideal, or a simple wire mesh hopper through which the birds can peck is another method. Supplies of greenstuff can come from the house, the local greengrocers, pick-your-own farm, or good restaurant. Hens don't like raw potato peelings or pea or bean husks, but all brassicas, lettuce, swedes, cooked onion and cooked potato, apple, pears, bread are good - here is your opportunity to spoil your birds as they do not have access to an outside area. Remember the nettle as a good source of insects and minerals.

**Daily Routine**

1. Check to see if all the birds are healthy.
2. Check food hoppers and fill up direct from 25kg bag.
3. Check water font or drinkers. Clean out any manure or litter in the drinker and fill up with fresh water if not using the low pressure system.
4. Collect eggs including any that have not been laid in the nestboxes.
5. Check the ventilation: if a windy night, shut down the ventilation and in the summer open it up. The building must never appear stuffy or smelly, neither must there be a howling gale blowing through.

An old fashioned hen house suitable for 12 birds. These houses are good for keeping hens in high rainfall areas so that the birds have somewhere to spend their day if the weather is wet.

Hen house as described in "Poultry House Construction" with a covered dusting box nearby.

A "Standard 10" hen house suitable for 6-10 hens.

A "Half Pint" hen house with roosting and laying area upstairs.

A small flock of Black Rocks and Warrens kept for eggs for sale at the garden gate.

Black Rocks at 17 weeks old. The birds look rather beaky at this stage, having just arrived from the rearer.

# GENERAL MANAGEMENT

This chapter contains useful do's and don'ts which will help you overcome several problems should you come up against them, and also increase the production of eggs from your birds.

1. What to do when your birds arrive; mixing birds
2. Handling
3. Treating for lice and mites
4. Internal parasites (worms)
5. Positive signs of health
6. Soft-shelled eggs
7. Broodiness
8. Which hens are laying?
9. Disposing of old birds
10. Killing
11. Feather pecking and egg eating
12. Moulting (full)
13. Moulting (neck)
14. Debeaking and beak trimming
15. Fox proof fence
16. Wing clipping
17. Crock eggs
18. Lighting
19. Housing
20. Feeding
21. Sexing
22. Vermin
23. Poisonous plants
24. Vaccinations

### 1. What to do when your birds arrive

Assuming that you have your housing all set up, make sure the birds have access to water - birds dehydrate very quickly. In the case of the house and permanent run method, the birds should be locked in the house overnight with food and water, to learn about their new surroundings, perches, nestboxes, etc. Then just open the pop hole; if they find their own way out they will find their own way back again.

### Mixing Birds

It is never good policy to add younger birds to an old flock, but if you have to, do so at night in the dark. Take the birds out of their crate and place them on the perches, ensure that there are extra hoppers and drinkers as the old birds are terrible bullies and will push the introduced birds away from food and water. It may take several weeks before the pecking order is established, if at all, and occasionally a bird mopes and starves to death.

2. **Handling**

Most hens and bantams kept as pets get handled a lot and they do not mind being picked up and cuddled. The correct way to hold any hen is with her weight being taken on the palm and forearm of the left hand and her legs gripped between the fingers of the left hand. This leaves the right hand free to examine the bird for lice which are under the tail if they are anywhere, and means the bird is balanced and comfortable. Even a bird not used to being handled will sit quietly with this method. The most practical way of handling your birds is either to catch them in their house or outside with a landing net.

3. **Treating for lice and mites**

Even though you have been putting louse powder in the nestbox and the dustbath, some hens will still have lice (sometimes called fleas). The cockerel must not be forgotten, either. It is important to delouse your broody before she sits on hatching eggs as it may put her off if she is covered with lice. The common chicken louse is host specific, ie. it only lives on chickens. It is not normally life-threatening to the chicken but the birds are better off without extra passengers. It can be easier to catch the birds at night if they are not used to being handled, but most louse powders do not recommend their use just before the birds perch for the night. Having caught your bird, hold it in the normal way (see above) then, still holding the legs, turn the bird onto its back on the ground or on a table. Use the weight of your forearm to hold the bird down so, in effect, your left elbow is about level with its head. Lice live all their lives on the birds so they lay their eggs on them too. These can be seen on the base of feathers looking like uneven lumps of white sugar. These lumps must be removed, so just give the feather a sharp tug and it will come out (it will regrow quite quickly) and then dispose of the louse eggs safely because if you just chuck them on the ground they will hatch and jump onto the nearest passing hen.

With your free right hand, sprinkle the louse powder under the bird's tail, under each wing and along its belly and rub well in. In severe cases you may have to rub some in along its back as well.

4. **Internal Parasites**

Usually referred to as worms. There are six different types inhabiting different parts of the hen, most of them in various areas of the intestine. Hens get an immunity to worms eventually but stress can disturb the hen's immune system and the worms then breed wildly and affect the health of the hen. Any bird off its food or moping is suspect. The preparation to use is

Another method is to pick up the bird in one hand so that the head is hanging down, slightly arched away from the body, and give it a quick, hard blow on the back of the head with a poker or similar instrument. If it is a young bird the skull is quite thin and it will be killed instantaneously, but will still flap and jerk as in the previous method. Normally two blows ensures an older bird is killed, but it can be a little messy this way, so always kill your birds on grass or over the garden.

A method of killing older and larger birds, particularly geese, turkeys and ducks, is with an airgun pellet through the back of the head. Hold the bird by its legs in one hand with its head on the floor so that its lower bill is resting on the ground. Put the muzzle of the airgun to the back of the head and pull the trigger. It is quick and perhaps better than the pulling of necks against broomsticks with larger birds, as this latter method can be difficult for the inexperienced poultry keeper. The airgun method is perfectly safe.

Always kill your birds out of sight of the rest of the flock.

### 11. Feather pecking and Egg eating

Egg eating is a common problem which we all encounter from time to time and normally it is started by soft shelled eggs, broken eggs in the nest box (insufficient litter) and/or boredom and stress. The old remedies of putting down mustard filled eggs, even curry filled eggs, normally do not deter the most determined egg eater. it is nearly always one hen that is the culprit and, of course, the others follow and join in the game. The most effective method of stopping egg eating and feather pecking is bitting your birds. Bits are small plastic 'C' shaped items which slide between the mandibles and clip into the bottom of the nostril. The bird can freely eat, drink and breathe, but can't close its beak totally. Pop these on for a month or two to stop the problem.

12. **Moulting (full)**

It is a natural part of a bird's life to change its feathers once a year, normally in August/September. Hens moult faster if they are not fat. All birds which only shed a few of their feathers naturally should be culled as they are poor layers; good layers go almost naked when moulting and tend to look rather scrawny. False or induced moulting in July for 18 month old birds is quite acceptable: feed them only oats and within a fortnight the birds will start to shed their feathers. When the feathers start dropping like autumn leaves gently increase their food ration to help them make new feathers quickly and strongly. Plenty of greenstuff is most important at this stage.

13. **Moulting (neck)**

This is a form of half moult due to stress and is more often found in cock birds because they have starved themselves in order that their hens can feed well, or after being badly frightened by a fox or dog. Watch the feeding of the birds and increase rations so that they all have enough food. We have seen birds with naked necks which were not in a neck moult - they had been stretching their necks through the wire netting to get at greenstuff the other side.

14. **Debeaking and beak trimming**

This is an operation which removes a small part of the upper mandible and it is carried out by a special form of electric pliers which cauterise the blood vessels in the beak. The operation, if done by a professional, is relatively painless. It is used in intensive systems where there is restricted space for the birds. It can be useful to trim hens' beaks if they have known freedom and you introduce them into a deep litter system, say in the depths of winter. The beak is made of the same kind of material as toenails and, like toenails, it grows, so if you bring your birds into deep litter for the winter and trim them, they won't be disfigured for long. Cut the top mandible back (with sharp scissors) so it is just shorter than the bottom one.

## 15. Fox proof fence

### Three styles of fox proof fencing

1. 8’ wooden posts with 6’ by 2” mesh wire netting and electric wire near base.

2. 2” x 2” angle iron with 6’ by 2” mesh on overhang and down to meet 4’ by 1” mesh on ground and up the side.

3. 8’ or 9’ wooden posts with 3’ by 2” mesh on top then 4’ by 2” mesh down to meet 4’ by 1” on ground and up the side.

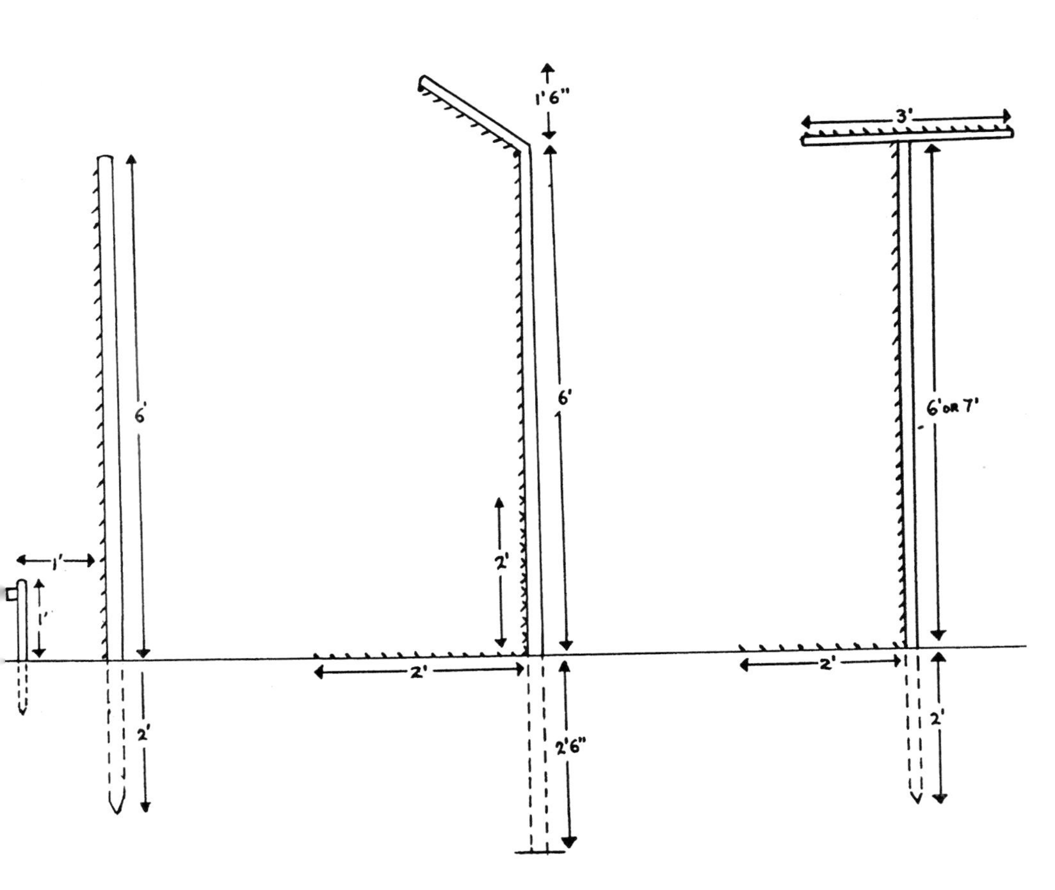

16. **Wing clipping**

When you have birds kept under method 3, there could be a problem with them flying out or trying to roost in the trees. Take a pair of kitchen scissors and cut the primary feathers on one wing only - the idea is to make flight impossible through imbalance, so don't clip both wings. Cut about $2^1/2$" from where the quills go into the flesh. This is important as if you cut too near the flesh the quills will bleed (difficult to stem the flow) or if you cut too far away, leaving too much length of feather this will render the clipping ineffective. Follow the line of small feathers for the correct place to clip. *See diagram.*

17. **Crock Eggs**

These come in various types and sizes, hen size and bantam size, china and plastic. One or two of these in the next box area will encourage the birds to lay in the right place, and to start to lay. They do need cleaning occasionally.

18. **Lighting**

This paragraph applies to those of you who opt for Method 4, page 16. In order to keep egg production at a healthy level, artificial lighting during the winter months is a must. A bird is stimulated to lay eggs by the length of daylight hours, and by increasing the daylight hours it has more time to feed and drink. The optimum time is 14 hours daylight and this is best done with a time switch that comes on early in the morning. This time switch needs to be adjusted as the days become longer or shorter. Adjust at the beginning of the month and in the middle of the month until there are fourteen hours daylight from sun up to sun down. By giving the birds artificial light in the morning (the light might come on as early as 2 am in winter) the birds do have the natural dusk to go onto the perches; if you had lighting in the evening a dimmer switch would be necessary otherwise the lights would go out and the birds would be all over the place. A low wattage bulb is needed, 40 watts is adequate in a 6' x 8' house. Under no circumstances change the lighting around too much - half an hour's difference at the most - otherwise you will induce a moult and you won't have a single egg.

19. **Housing**

What to look for when considering a new poultry house: is there sufficient ventilation with no draughts? Are the nestboxes easy to get at from the outside of the house for egg collection? Are they sited in the darkest part of the house as the hen will want to lay in the most secret dark place she can find? Are there perches, and are they wide enough (2" and at least 8" of length per bird)? Are they the right height so the birds use these and not the nest boxes to perch in? Is there a droppings board to make cleaning out easy and to keep the floor of the house cleaner? Depending on the system you use, is the house easy to move? Is it fox-proof? Has the timber been treated, preferably by a vacuum process such as tanalising? Is there felt on the roof to harbour red mites? Has the house been designed by people who keep chickens and therefore understand their requirements? Is the house substantially built so your investment will last for years?

20. **Feeding**

If you can find an additive-free feed you are fortunate, but the trend is going that way. The ingredients of all feedstuffs ought to be declared on the bag, so people can choose, but very few mills will tell you which ingredients they use; they talk in terms of protein percentage but it is the **type** of protein which is important. Feathers are protein and are used in some poultry feeds. Fishmeal is the finest and best value protein as the hen can use it all and therefore needs less of it. Grain in the form of wheat is a good basic diet but more protein needs to be added for egg production. It is possible to get feed made up to your own formula if enough people get together to use it within its shelflife. You are not allowed to feed meat to poultry unless licensed under the regulations.

**Water**

Vital for all life processes - sometimes called the forgotten nutrient.

21. **Sexing**

Professionals who sex chicks at day old usually do so with hybrid hens but not with pure breeds. It is a great deal easier, and more accurate, to wait for the feathers to grow: when the males are about 6-9 weeks old they will grow pointed feathers on their lower backs, the saddle hackles; females have round ended feathers here.

22. **Vermin**

Where there is livestock, vermin will sooner or later be there too, whether it is foxes, rats, mice, wild birds, flies, mink, lice, mites, etc. Prevention is the secret and this means storing feed in dustbins so that mice and rats are not attracted by the smell, and picking troughs up at night and putting them in the house with the hens (if they still have food in them). We have written "Modern Vermin Control" (£7.00, inc. postage and packing, from Domestic Fowl Research) which covers this subject in depth.

23. **Poisonous Plants**

There are certain poisonous plants commonly found in gardens. These include yew, laurel, privet, laburnum, aconitum (monkshood), euphorbia, lords and ladies (arum), datura and foxglove. Bear in mind that most poisonous plants taste nasty so hens will rarely be affected.

24. **Vaccinations**

Today, there are four main vaccinations which are used. These are not always necessary and some people do not vaccinate at all.

a) *Mareks* - this is a particularly pernicious disease, striking down the young birds between 4 and 22 weeks of age. It is mainly borne by feather dust, in young birds and is very contagious, with high mortality. There is also congenital Mareks, which is often seen in very closely bred show birds such as Sebrights, Silkies, Hamburgs. This is passed through the egg and will not affect other breeds or stock.

The chicks are vaccinated (MAREXINE made by Intervet, Cambridge) at day old, either in the neck, wing or thigh - the vaccine must be used within 3 hours and it is not expensive.

b) *IB, Gumboro and Newcastle Disease* - IB stands for Infectious Bursitis: Gumboro and Newcastle Disease vaccinations are all administered to chicks via the drinking system or drinker. This is a very simple task and again not expensive.

# HATCHING

The time will come when you want to hatch your own chicks. You have several choices:

1. under a broody chicken with a) your own eggs
   b) bought-in eggs
   c) bought-in dayolds
2. in an incubator with a) your own eggs
   b) bought-in eggs
3. under a brooder with bought-in dayolds

Care of the eggs is the same whichever method you choose.

Storage should be blunt end up in a temperature of 50° F (10° C) for up to 7 days, or turned end over end if stored longer; or prop up first one side then the other of the egg trays at about 30°. Storage after 10 days reduces hatchability. If the eggs are very dirty do not set them; small amounts of dirt can be removed by sandpaper. Washing eggs reduces hatchability, but if considered necessary, use water warmer than the eggs: this is to ensure that the membrane inside the shell expands on contact with the warmer water, helping to exclude bacteria. If you use water colder than the egg the membrane shrinks and draws any bacteria in through the shell which, of course, is porous. Use Virkon S as a cleaning agent. Choose only good sized, normal shaped eggs, not less than 24 hours old, from birds which have been laying for at least 2 months if they are young birds. It is important to realise that eggs go into a dormant state after they are laid and do not have to be kept warm. They remain dormant until they meet the correct conditions for growing - like a vegetable seed. "Setting" the eggs is either putting them under a broody or in an incubator.

1. **Under a broody hen**

This is the traditional way and often gives the best results, as long as you have enough hens broody at the right time.

Broody boxes are best constructed as per the diagram, see page 24, and with wire netting on the base to prevent rats from burrowing into the nest. They are best set directly onto the earth which allows beneficial natural moisture to come up through the nest. The boxes can be made in banks of any number but four or five makes for easy handling before and after the hatching season. The boxes should be set on a small mound, about two turves high in case there is a lot of rain, and in the shade. They must not get too hot as this is likely to put the broody off. Punch a shallow dip like a saucer in a turf and lay this, grass down, in the box. Line the depression thinly with sweet hay or straw. Make sure the turf fits well so that no eggs can get rolled out into the cold. Put crock eggs or marked fresh eggs into the nest ready for the broody; she needs to sit steady on these for a few days before you put in the eggs you want her to hatch.

It is most important that she is separated from the rest of the hens as, not only will the cockerel disturb her, but the other hens will lay with her eggs and they will not all hatch together.

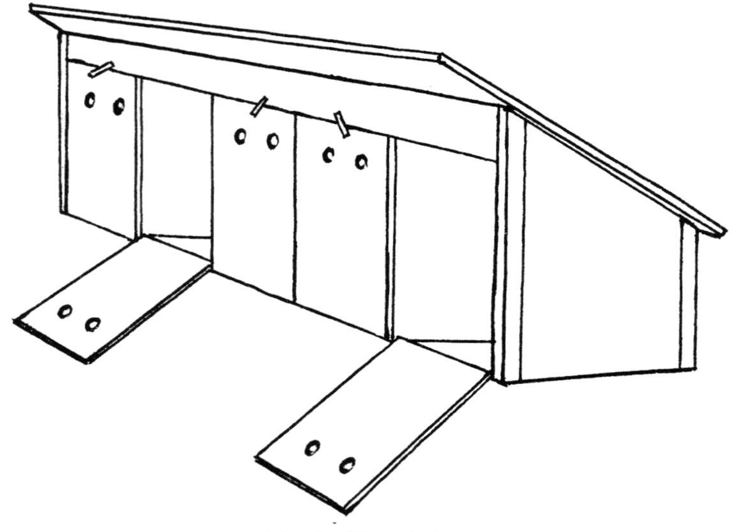

*A bank of broody boxes*

Select your broody from as large a breed as possible, such as Sussex, Cochin, Orpington, Rhode Island Red, as using Silkies or Silkie crosses will limit the number of eggs you can set due to the small size of the bird. Although some hybrid hens have been known to sit well, they are on the whole totally unreliable. Broodies will not want to leave their nest if they are serious and will guard any eggs jealously, fluffing up their feathers and grumbling at you, even pecking at intruding hands. In order to check that your selected broody is serious, take any eggs out and slide your hand under, palm up. She should "cuddle" your hand with her wings. You must delouse any broody with louse powder before getting her to sit for you as she will otherwise be irritated and disturbed by fleas. If your broodies are not used to being handled, it is best to put them in a cardboard box with straw and eggs and close the lid. You will then be able to transport them to the broody boxes easily. Let them sit in the cardboard box to regain their composure for an hour or two and then you can pop them quietly into the broody box onto the crock eggs already here. Alternatively move them when it is nearly dark.

Unless you have only one or two broodies, it is best for you to get them off the nest every day. Try to do this at the same time and tether them far enough apart so that they cannot fight. To tether, use a thong or piece of string attached to the broody's leg with a sliding loop and, on the other end, a curtain ring dropped over a three foot high stick. Water and whole wheat only must be within reach and the birds should be off the nest for about 20 minutes. Check that each one has defecated before you gently put her back. If not, you may have to help this procedure by raising the hen to waist height and then dropping her on the ground. Several times may be necessary. As you are putting her back on the nest, check that her feet are clean - hen manure will easily turn the eggs bad. If you want to use a system of putting the broodies in individual wire cages when taken off the nest, it is an alternative to tethering them but involves more equipment.

When the broodies have sat for a few days on the crock eggs and become accustomed to whatever method you use for getting them off the nest each day, it is the time to set the eggs you want hatched.

Put the eggs under her in the evening - remember to take out the crock eggs - and after about an hour check she is covering them all properly. Take away any which are not covered. An odd number of eggs fits into a circle.

A good broody will stay broody until she hatches off some youngsters. This may be after 21 days or three months if you are juggling the eggs around to make the best use of the broodies. Don't be afraid of keeping a broody sitting on crock eggs until you are ready to set some hatching eggs. When you do set some clutches, try to set two or three at the same time. This means you will be able to amalgamate clutches, after having discarded infertile or bad eggs, and start one of the broodies off on another clutch. If a broody does become fierce and insists on pecking you she is only trying to protect her eggs. Offer your hand with your palm uppermost where the thicker skin will withstand the pecks better, and turn your hand over once it is under the bird. Long sleeves are useful. Protective gloves are not really recommended as you can't feel what you are doing.

**How to candle eggs**

Candling was originally done with the light of a candle for checking on embryo growth, and is basically a concentrated source of light with which to see through the shell and into the eggs. Equipment available commercially ranges from a simple hand grip with a 10 watt bulb inside, to ultra violet "cool" lights and sophisticated machines which candle thousands of eggs automatically. A DIY version could consist of a three-sided wooden box with a household light fitting inside and a 1" hole drilled in one side.

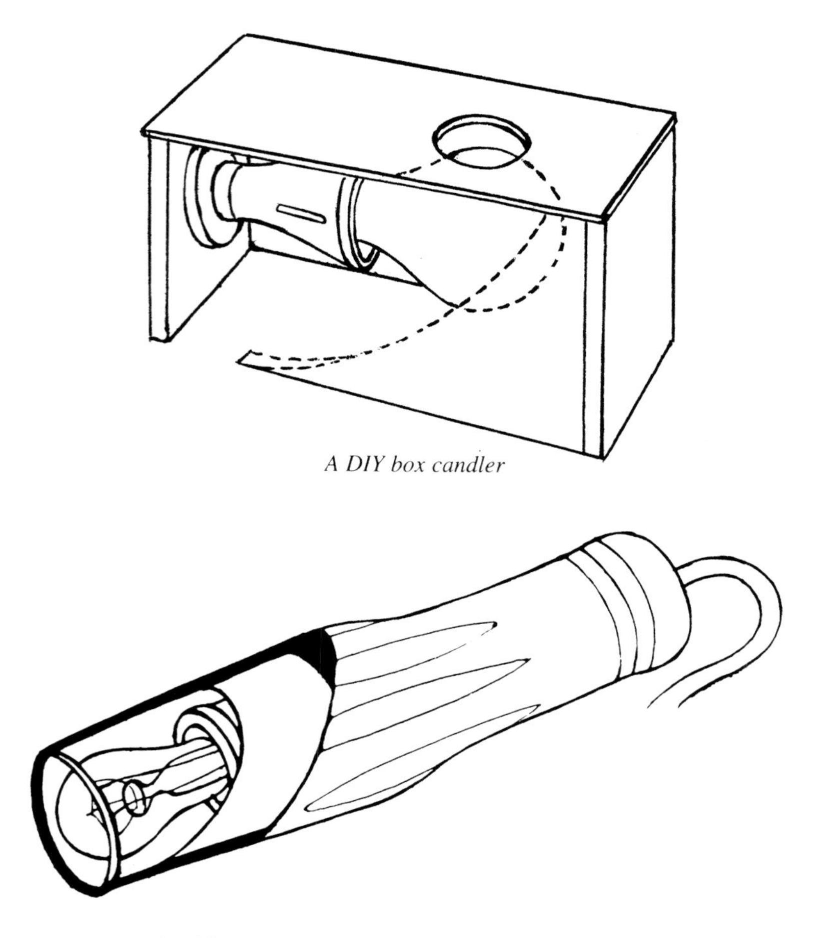

*A DIY box candler*

*Hand held candler, quicker to use then the box version*

The hand held version is more flexible. The broad end of the egg is held to the light making sure that the light is pointing away from your eyes and that the room is as dark as possible. The air sac will be visible as a circular line at the broad end. The various stages of incubation are illustrated below. The infertiles have no dark areas. Be especially careful if you can smell rotten eggs whether near the broody or in an incubator. If one explodes over you, you will be avoided by all humanity for a long time!

*Stages in incubation:*

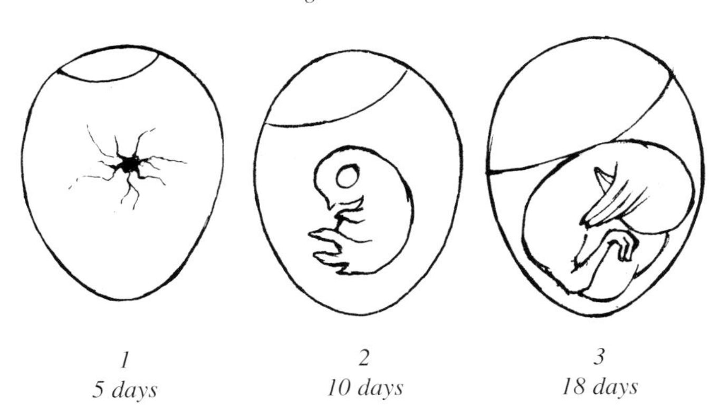


Your broody hen is now settled and used to you and the daily routine, and you have practised candling using some eating eggs. If you are using Method a), **your own eggs**, you will have chosen your breeding pen of birds and if they have been running with the wrong breed of cockerel you will have waited two weeks before starting to collect and store eggs to ensure they are pure. Check that the cockerel does not have fleas (under the tail) as that can affect his fertility, and also that he not too old - less than 3 years preferably. Look at his spurs on his lower leg - very small = up to 6 months old, 1/2" - 3/4" = 1-2 years old, over 1" = any age up to 10 years old!

Put the required number of stored eggs under your broody, remembering to take out the crock eggs or the marked eggs you were using to keep her broody. Make a note in your diary of the date. It does not matter what time of day you set the eggs but your broody may be quieter in the evening in the dark: do not shine the torch near her, however, as this will frighten her badly: Talk to her to keep her calm.

Method b), **bought-in eggs**, will have to be used if you do not keep a cockerel. Try and get them from a reliable source so that they are fresh and the breed you want. You may well have to order them in advance. No-one can guarantee the fertility of the eggs, but if your source is hatching from the same birds they will know if the eggs are good or not. Remember that hatching eggs need care when being transported, and let them rest for 12 hours before you set them under the hen. Bantam eggs are very much smaller than hen eggs and should not really be set under the same hen with hen eggs. Bantam eggs, depending on the breed, can hatch a day before hen eggs. In an incubator this would not matter, but a broody hen stays broody until she hears cheeping and you do not want her to leave the rest of the good eggs just because one has hatched early.

Method c), **bought-in dayolds**. You may want to buy in dayolds if all the eggs you set are no good for one reason or another. The dayolds must be just that - one day old. You cannot put three-day-old or week-old chicks under a hen and expect it to work; the hen is cleverer than that. All her maternal instincts are geared to teaching and protecting a freshly hatched chick and recognising it as her own (ie. one that she has hatched) and all the chick's instincts are geared to being guided and protected. Birds have colour vision, but the chicks do not have to be the same colour as each other, as the hen will accept any colour but ONLY in the first day - your own eggs may have come from different coloured hens and will hatch out different colours, for instance. The first day is rather like a photograph for the broody hen: if you add another colour chick on the third day, it will not fit in the original picture and she may kill it. You *can* fool her as she cannot count, by adding more of the colours already there, if only one or two of your own eggs hatch. The maternal instinct is very strong and anything which does not fit in to that critical first day picture is a threat. You will see this in the way a tiny hen will try and protect her chicks from anything she considers a threat whether it is as big as a cow, the cat, another hen or someone else's chick.

Having decided to buy some dayolds you will have contacted the hatchery to order them and you will have discovered which day of the week is hatching day so you can arrange to collect the chicks (take a small box with shavings) or have them sent by road. They still have the yolk of egg inside them to act as a food source for 48 hours after they hatch. When you get them home **don't** put them straight under the hen. They will be thirsty, so pick each chick up and dip its beak into water to make sure it has a drink. When all the chicks have had a drink - don't let them get cold - take the box to the hen. This is to trigger her aggressive broody or sitting instinct into her protective maternal instinct and you will hear by her change of tone that this has happened. Then quietly and slowly put the chicks under her and take the remaining eggs out at the same time - chick under, egg out. The chicks will cuddle under the warmth of the hen and then keep quiet. If they had actually hatched under her they would have started their cheeping still inside the egg to trigger her maternal instinct. If she does not hear them cheeping when you put them under she may attack them as her broody aggressive instinct will still be dominant, not her maternal one. Cheeping is a vital link between the hen and chick as it ensures survival of the chick and keeps the maternal instinct strong.

It is not a good idea to mix large fowl and bantam chicks as, due to the difference in size, the bantam chicks can get bullied and pushed off the food: the chick of a Dutch bantam can be 1/4 the size of a Light Sussex chick. Having successfully put all the chicks under your hen, leave her in peace and quiet. She will be purring. After a while she will encourage her chicks to eat and drink by a different call and you can then leave the main task of rearing them to her, merely providing food, water and shelter.

Method 1. a) Returning to **your own eggs** under your broody, you will have candled them possibly twice during the three weeks, taking out the infertile eggs. If you are not sure if an egg is infertile, just leave it with the hen and get someone who has some experience at candling to show you what to look for. Assuming your chicks have started to hatch at the correct time, ie. 20-21 days after you set them, do resist the urge to pick the hen up and have a look underneath. She really does not need disturbing at this stage. A small pan of chick crumbs should be placed close to the broody and a suitably small drinker also. The broody may not leave the nest for two days while she waits for the chicks to hatch. This is perfectly normal. If she has not left the nest after two days from the time you know the first chick hatched, it may be that the remaining eggs are not going to hatch. Gently, talking to her all the time, slide your hand under her and extract any eggs you can feel. Shake these gently beside your ear. If they rattle they are no good and will not hatch, so carefully dispose of them as they may be rotten. Then you can move the birds to a rearer coop and run.

1. In an incubator with a) **your own eggs**

Follow the same storage conditions as 1. a) and when you have got enough eggs you can set them in the incubator. It is difficult to generalise about incubators and their usage as the various models differ. It is imperative that the manufacturer's instructions are followed closely. Incubation is a complete science on its own, everyone achieving different levels of success with seemingly identical equipment. It is important to stress that an incubator is only as good as the person operating it.

The main advantage of incubators over broodies is that they are available at any time of the year and they don't need feeding while they are not incubating.

What most manufacturers do not tell you is the importance of correct cleaning between hatches. Virkon S is the stuff to use, as the object is to sterilise. More poor hatches are caused by bacteria left over from the previous hatch than by any other cause.

Method 2. b) **bought-in eggs**. The same principles apply as to bought-in eggs under a broody concerning reliability of source and care in transport. Don't forget to let the eggs rest for 12 hours before you set them, and, of course, you have had the incubator running for two to three days to check that everything is working correctly.

Method 3. a) **under a brooder with bought-in dayolds**. For small numbers, a large cardboard box with the brooder at one end is ideal. Having ordered and collected, or had sent, your dayolds, the first action is to give them a drink by dipping the beak of each chick in water. Then place each chick under the brooder so that it knows where the warmth is. A dark source of heat is best so that the chicks can sleep at night. They seem to grow better and feather-up better if they have natural darkness, so avoid heat lamps which give off light, even if they are red in colour. The electric hen has been around for a great many years in various forms and is ideal, as the under surface is soft and yielding as well as being warm - as close to a broody hen as you can get. Start off with a flat pan for the chick crumbs so the chicks can find the food easily and a proprietory drinker, preferably with a red base to which colour they are attracted. Their instincts for feeding and drinking are very strong - they do not need a broody hen to teach them - but if we can help them in the early stages they will grow on that much better.

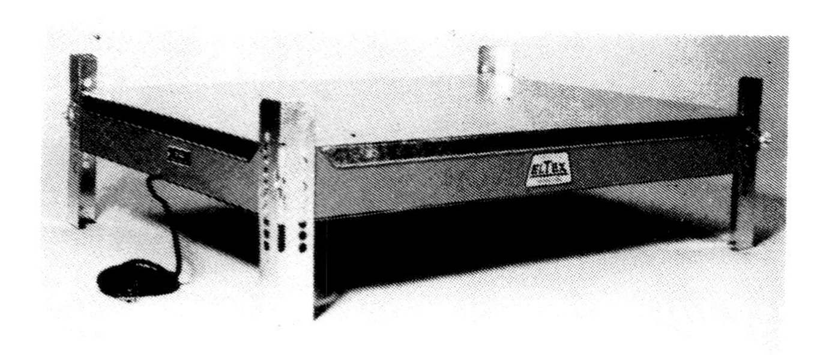

*Electric hen*

# REARING

1. **Under a hen**

Once the youngsters are dry and strong enough they should be moved from the broody box to the rearing coop and run.

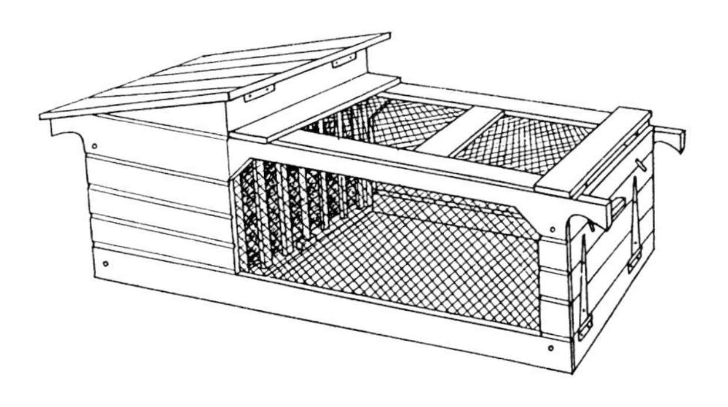

**SIZE 5' X 2' X 2' high**
**(152 x 61 x 61cm)**

**BROODY COOP**

This comes with a sliding roof, lift-out top to the run section and a drop-down flap to allow outside access. A wire mesh floor is available as extra.

Place the coop and run as close to your house as possible, on the lawn, and try and keep dogs and cats away. The coop should have a solid floor with straw on it. The run must be covered on the top with wire netting to prevent predation by crows and magpies, as well as on the sides and base to stop rats burrowing in. If it is rainy, drawing-pin some heavy gauge polythene on the top and the side of the prevailing wind. The hen and the chicks will move freely between the coop and the run but, to begin with, put the food and water close to the coop end. Remember that your broody is now extremely hungry and thirsty. The first thing she will do when you put her in the run is to make a huge mess. Remove this and give her some chick crumbs in a shallow pan. She will automatically call the youngsters to the food and show them the drinker. The youngsters may not be hungry the first day if the yolk sac has not been fully absorbed. The hen will scratch the chick crumbs everywhere. The water must be in a drinker or font and never in an open dish otherwise the young will get chilled and may drown themselves. After a few days, change to a roll top feeder, or small plastic feeder, so they can neither scratch the food around nor mess in it. Remember to move the coop to fresh ground every day.

1. **Under a brooder or artificial heat source**

The chicks will soon progress to a roll top feeder or plastic feeder which will prevent them scratching food all over the floor. They will need cleaning out regularly - a dustpan is about the most efficient tool for this if the chicks are in a box. If you cannot obtain an electric hen type brooder for rearing the chicks, a ceramic infra-red lamp is the next best thing, ie. it has heat but no light. If your rearing area has no windows you will have to provide light during normal daylight hours. The chicks will need some form of heat until they get their feathers which, depending on the time of year, can be 4 - 7 weeks. The temperature 2" above the shavings should start at 95$^{o}$F (35$^{o}$C). Check it with a thermometer. The chicks will soon tell you if it is not right by huddling together and cheeping loudly if they are too cold or scattering to the far edges of the box and panting if they are too hot. The heat can be gradually reduced during the rearing period by raising the heat source or fitting a thermostat and setting it at a lower temperature. As the chicks grow they will need more space as they will not only be bigger but will be spending longer periods away from the heat source. Extend the rearing area by changing to a much bigger box or change to a ring of hardboard made from two strips 6' or 8' x 18", fastened with bulldog clips at the top and a brick on the floor. The bases of the drinkers will become filled with shavings as the chicks are always scratching about. Raise the drinker by standing it on an upturned feeding pan, or a spare drinker base. Check that it is not so high that the chicks cannot reach the water. The roll top feeders will get a certain amount of shavings in them too, so when you fill them up tip the remaining chick crumbs down to one end, scrape off the shavings and then fill with fresh food. This way the food does not become stale. Place the feeder onto a weld mesh stand.

**Both systems of rearing.** The chicks should stay on ad lib chick crumbs for at least 6 weeks and then go on to grower's pellets. They can be started on wheat at about 8-10 weeks but mixed grit must be given at the same time.

**How soon can you integrate your youngsters with the older birds?** If they are with a broody hen they can be introduced at about 8 weeks as she will protect them from other hens, but do watch that they do not get bullied. If they are not with a broody they cannot be introduced until the youngsters are full-grown - about 6 months old - otherwise they will get bullied and pecked and possibly killed by the older hens. These birds will be defending their territory and their pecking order, so until the youngsters can stand up for themselves they are at risk.

**Galvanised or Plastic?** The galvanised drinkers will last a lifetime. Sizes 2 gallon, 1 gallon, 1/2 gallon.

The Plastic drinkers and feeders will last several years but are 1/4 of the cost of the galvanised drinkers. Sizes 3 litres and 1 litre.

**Plastic Bantam Feeders** Very useful, keeps the food together and dry.

**Rearing from 6 weeks with no broody.** If the weather is reasonably warm you can start to turn off the heat during the day in order to harden the chicks off before you put them outside. They should have had no heat at all for at least 5 days (nights as well) before they go out into a small house and run. They can be allowed to range free, as long as they have their own sleeping quarters, and do make sure that you reduce the size of the pop hole or entrance to the run so the older hens cannot get in and eat their food or bully them. As the growers (they are no longer chicks) have been used to natural light they will adapt very readily to going outside. If birds have been reared on continuous light they will panic when it gets dark and crowd into a corner and suffocate each other. A wire or slatted floor in the rearing house will enable the ones underneath to breathe if they do crowd together.

**Electric poultry netting.** The close mesh 42" high plastic netting recently developed is useful for keeping young stock separate from the older ones even if it is not electrified as the mesh is too close to let the older ones through. When electrified it not only keeps the older ones in without wing clipping but keeps the fox out and can easily be moved onto fresh ground.

**What do you do with your surplus cockerels?** Everyone who breeds poultry has this problem. You may be able to give them away, you may be lucky and find a breeder who wants one pure bred cockerel. The cockerels will not only eventually fight amongst themselves but the resident male will not appreciate competition and will do his best to suppress it. You may be in an area where there is a local poultry market (contact MAFF for details, see list of useful addresses), or if you are keen on growing your own, eat them. If surplus cockerels have been regarded from the early stages as destined for the table, it is much easier - at least you know what they have been fed on.

# DISEASES AND AILMENTS

SPEED is essential.
A *post mortem* will establish cause of death and possible treatment of other birds.
ONE medicine at any one time.
Most medicines are only obtainable through a vet.
Wash hands after handling medicines.
Do not eat eggs or birds when medicines are being given nor for a week afterwards, or follow the withdrawal instructions on the label.

| SYMPTOMS | NAME | CAUSE | TREATMENT | BIRD *Species Affected* |
|---|---|---|---|---|
| Listless, head sunk into neck. White diarrohea, sometimes blood in droppings. | Coccidiosis | Coccidia parasite. | Add Proleth to drinking water. Make no other water available. | Chickens, Turkeys, Peafowl, Pheasants, Guinea Fowl, Quail |
| Listless, head sunk into neck. Yellow diarrhoea. | Blackhead | Parasite carried by Heterakis worm. | Add Salfuride to drinking water. Renew after 3 days if not all consumed. Make no other water available. | Chickens, Turkeys, Peafowl, Pheasants, Quail, Guinea Fowl |
| Listless, green diarrohea. Gaping in pheasants. Loss of weight in hens and waterfowl. | Worms | Up to 6 different species of worm inhabiting different internal parts. | Flubenvet mixed into feed in trough. 1/2 teaspoon to 2lb feed. Geese and ducks, 3 days, all else 7 days. | Chickens, Ducks, Geese, Turkeys, Peafowl, Quail, Guinea Fowl |
| Excessive scratching, visible fleas around vent. Redness around vent. Sometimes colonies of eggs on base of feathers. White dust around perch fittings - mites. Lassitude. Infertility. | Fleas or mites | 4 types of flea. 2 types of mite. | Dust with flea powder particularly around vent and under wings. About 1oz per bird. Spray housing with Duramitex, especially around perch fixings where red and white mite live and feed on birds at night. Put flea powder in nestboxes. | All birds |
| Sneezing, discharge from nostrils, foam in corner of eye. Swollen sinus in turkeys. Sweet, sickly smell. | Mycoplasma (cold) | Bacteria. | One injection of 1/2 ml. of Tylan 200 per bird in breast muscle. 1ml. for very large birds, eg. turkeys. Repeat only in very bad cases 3 days after first injection. Mild cases, Tylan soluble, or Terramycin, in water. | Chickens, Turkeys, Peafowl, Pheasants |

| SYMPTOMS | NAME | CAUSE | TREATMENT | BIRD *Species Affected* |
|---|---|---|---|---|
| Blood | Wound | Feather-pecking due to over-crowding. Accidental cut. | Terramycin Gentian Violet spray on area and isolate until healed. In bad cases 'bit' all the birds for 2 months. | Young Stock, or any other birds. |
| Raised, encrusted scales on legs. | Scaly Leg | Mite, burrowing under leg scales. | Use surgical spirit or Quellada from Boots. *See page 24.* Do not pull off crusts. | Any bird. |
| Brown diarrohea, slow growth, poor feathering, in young stock, affected by the cold, several deaths overnight. | Enteritis | Escherichia coli (E. coli) bacteria, usually due to poor ventilation, dirty conditions. | 1 measure of Terramycin per gall. water changed every 24 hours for 3 days. If that doesn't work (bacteria immune?) use Framomycin. I measure per gall. water for 5 days. | Young stock of Chickens, Ducks, Geese, Turkeys, Peafowl, Pheasants, Quail |
| Purple comb, when normally bright red. | Heart disease | Age or deformity. | No treatment. | Chickens |
| Round swelling on underside of foot or ankle. | Bumble foot | Injury first, then infection. | There is no treatment other than a jab of Penicillin. The foot/ankle will always remain swollen. | Chickens, Pheasants, Turkeys, Guinea Fowl |
| Top part of beak grown much longer than bottom. | Overgrown beak | Deformity - beak should meet exactly. | Trim back to level with lower portion with sharp scissors. Be careful not to cut the quick. | All birds |
| Sides of females bleeding or bare of feathers. | Bareback | Sharp, long spurs of males. | Trim cockerel's spurs with hacksaw, being careful not to cut the quick. File smooth and rounded. | All birds which have spurs |
| Unusual behaviour. | Stress | Unusual disturbance or major changes - new stock added. | Protexin is ideal. Comes in soluble or in-feed form. | Any bird. Stress can cause dormant illness to take over the body. |
| Wasting away but still feeding. | Avain tuberculosis | Bacteria. | No treatment - carried by wild birds. Natural immunity possible. | All birds |

# USEFUL ADDRESSES

## THE NATIONAL POULTRY CLUB

c/o Michael Clark, 30 Grosvenor Road
Frampton, Near Boston, Lincs TE20 1DB

Tel: 01205 724081

## THE BRITISH WATERFOWL ASSOCIATION

Gill Cottage, New Gill
Bishopsdale, Leyburn, North Yorks

Tel: 01969 663693

## THE NATIONAL FEDERATION OF POULTRY CLUBS

c/o Harold Critchlow
Little Longsdon Farm, Longsdon
Stoke on Trent, Staffordshire

## FANCY FOWL PUBLICATION LTD

c/o Scribblers Publishing Ltd
The Watermill
Southwell Road
Kirklington
Notts
NG22 8NQ

# INDEX

# A LIST OF WEEKLY POULTRY AUCTIONS IN THE U.K.

| COUNTY | CITY/TOWN | DAY | AUCTIONEERS | TELEPHONE NO. |
|---|---|---|---|---|
| Cheshire | Chelford | Monday | F. Marshall & Co | 01625 861122 |
| Cornwall | Liskeard | Thursday | Kivells | 01579 346938 |
| Devon | Bideford | Tuesday | Kivells | 01237 472146 |
| Devon | Hatherleigh | Tuesday | Vicks | 01837810496 |
| Devon | Holsworthy | Wednesday | Kivells | 01409 253465 |
| Devon | Newton Abbot | Wednesday | Rendells | 01626 353881 |
| Essex | Colchester | Saturday | Essex & Suffolk Auctions | 01206 842156 |
| Herefordshire | Ross on Wye | Friday | Williams & Watkins | 01989 762225 |
| Kent | Sevenoaks | Monday | Pattulo & Partners | 01732 452329 |
| Lancashire | Clitheroe | Wednesday | Clitheroe Auction Mart | 01200 423325 |
| Leicestershire | Melton Mowbray | Tuesday | Melton Mowbray Mkt | 01664 562971 |
| Middlesex | Southall | Tuesday | Richard J. Steel | 0181 5741611 |
| Norfolk | Norwich | Saturday | Homors | 01493 750225 |
| Shropshire | Bridgnorth | Monday | Nock Deighton | 01746 762666 |
| Staffordshire | Leek | Saturday* | Leek Auctions Ltd | 01538 372749 |
| Staffordshire | Uttoxeter | Wednesday | Bagshaws | 01889 562811 |
| Suffolk | Bury St Edmunds | Wednesday | Lacy Scott & Knight | 01284 763531 |
| Warwickshire | Henley in Arden | Wednesday | Henley Auction Sales | 01564 792154 |
| Wiltshire | Salisbury | Tuesday | Southern Counties Auctioneers | 01722 321215 |
| E. Yorkshire | Beverley | Wednesday | Beverley Livestock Centre | 01482 861800 |
| N. Yorkshire | Selby | Saturday | Selby Livestock Mart | 01757 703347 |
| S. Yorkshire | Penistone | Saturday | Wilbys | 01226 299221 |
| Lanarkshire | Lanark | Thursday+ | Lawrie & Symington | 01555 662281 |
| Roxburghshire | Newton St Boswells | Monday | John Swan & Sons | 01835 822214 |

There are no poultry auctions in Wales. * Fortnightly + Monthly

There is a number of specialised auctions and sales held by various organisations and auctioneers throughout the year. If you are going to use them, make sure that your birds are in show condition as many of these auctions have a show grading system to help the buyers. Others, sadly, have become dumping grounds for surplus stock, so it is worth choosing your breed carefully and also having a sensible breeding programme. I have seen Welsummers go for extraordinary prices one year because there were only three pens of well presented birds, and the next year there were pens and pens of Welsummers, all obviously hoping to catch the previous year's price and failing, mainly because of numbers and also because of poor quality and presentation. Most of the best breeders do not use auctions as they can sell their birds privately, thus saving the auctioneer's commission, but other people rely on auctions as a means of selling their birds. Caveat emptor! (Buyer beware).

# ADVICE ABOUT SELLING AND BUYING AT AUCTIONS

## IF YOU ARE SELLING:

A) Check to see if the auction is going ahead as scheduled; sometimes there are seasonal changes such as at Christmas when they might be selling only turkeys, and sometimes, although rarely, the market has had to be closed because of an outbreak of disease such as Newcastle Disease for instance.

B) Choose the birds you are going to sell and get them into a clean pen a day or so beforehand; (they will be easier to catch on the day if they are confined). Check them for lice and fleas if necessary, and clean them up, especially the legs and feet.

C) You must have suitable containers in which to transport your birds. Plastic or wooden crates are best, but anything is allowed as long as it is rigid and there are plenty of air holes; be careful though, that heads, feet or wings do not get trapped outside the container. A suitable trap door system must be arranged so that you can have access to the birds and they don't escape while you are unloading them. It is best to have a bed of shavings in the container as well.

D) Get to the auction early so that you can put your birds in their cages in good time, not too soon and not too late, that is if cages cannot be booked. Remember to arrive earlier still during school holidays, as weekly markets become very busy at these times.

E) Handle your birds with care across the body and wings and not just by the wings or legs, when putting them in or taking them out of the cages. Sometimes birds are most reluctant to come out of the cage after the auction, so you will need two hands to gently ease them out. If you are a little worried about this, try to find an experienced person to help you. The main thing is not to stress the birds even more after an already stressful day.

F) Make sure the birds have some water once they are in their cages. All auctions have to provide containers for this, but as some are more adequate than others, many people like to take their own.

G) If a bird is sick, injured or badly stressed it should be removed. It would not be a good advertisement for you, and of course a diseased bird must not be put up for sale.

H) Occasionally, when you have several birds in a cage together, one may become aggresive and start to attack the other birds. This would be mainly due to stress and could result in blood everywhere. Not a good advertisement, so remove the offending bird and cage it seperately if possible.

I) Don't mix breeds, ie have ducks and hens in the same crate, during transport.

**IF YOU ARE BUYING:**

A) Take suitable containers such as crates to put your purchases in.

B) Get there in plenty of time and buy a catalogue if necessary; these often have a nasty habit of being in very short supply on the day of the sale. You may also be lucky and meet the breeder of the birds you are after.

C) You might need a buyer's number from the auctioneer so obtain one of these if you do.

D) Look at all the birds you are interested in, checking for signs of age and to see if they are the right size and shape; buying in an auction is a bit of a lottery! Try to talk to the owner / breeder of the birds. Remember, no breeder is going to put his best birds in an auction, so in fact you are buying seconds or birds that don't come up to the breeder's expectations.

E) Set yourself a price for the birds and walk away if they go over. There is always another day, and besides, remember what is said in paragraph D above. I have seen birds make stupid prices because of ignorance and the "must have" syndrome.

F) When you have bought and paid for your birds, give them something to eat, (whole wheat is best), as they may have been travelling etc. for many hours.

G) Having crated or boxed your birds and put them in your vehicle, take great care if it is a hot day. I have seen birds die of heat exhaustion and dehydration just because the owner has stopped the car and gone to get a cup of tea, so when it is hot, always travel with plenty of ventilation and the air conditioning turned on if possible.

# SHOWING OR EXHIBITING YOUR BIRDS

This chapter has been added because it is an important adjunct to keeping chickens and bantams. Not everyone is interested or has the time to show their birds, but nearly everyone who keeps chickens enjoys seeing birds being exhibited, especially at agricultural shows where it is interesting to make comparisons with their own stock. So in this chapter we will try to inform you about the procedures for showing, although we will not go into too much detail because exhibiting is an art in itself, but there should be just enough information to ensure you some success.

For the serious exhibitor, showing involves a great deal more than catching up the birds the night before, cleaning them, putting them in a box and setting off early the next morning for some village hall, marquee or agricultural show. It is all about planning which should start at least a year before the beginning of the show season.

First of all choose the breed or breeds you like the look of and want to keep. A lot of people are drawn to the laced breeds such as Sebrights and Laced Wyandottes, but getting the right shape of bird and the correct feathering pattern is quite difficult, and you may well be competing against some very experienced breeders and exhibitors. So it is advisable to choose a single coloured bird, and of course the best place to see a good variety of breeds is at a Show.

When you have decided what you

*A Silver Laced Wyandotte Bantam in a Show Pen*

would like, approach one of the better known breeders of stock, but before you do this, arm yourself with a copy of the Poultry Standards Book so you can be aware of the defects that can arise in that breed. You are looking for size, shape and age, comb, eye colour, earlobe and leg colour, feathering and any deformities such as crooked beaks or toes. It is very hard sometimes for a novice to take note of all these points, but if the breeder is worth his salt, he won't sell you a duff bird. Having said all that, you must remember that he will not be selling you his best stock either, only his cast-offs, but you will be buying into his gene pool which may produce a winner. Some breeders are very fair and if you get on well with them, they might use you as a satellite breeding centre, particularly if the breed is rare or rather scarce.

*A good light Sussex (Large Fowl) cock in a show cage, note the up/down sliding cage door*

Males are easier for a beginner to show than females. There is quite an art to the timing and feeding involved in bringing a young female up to show size and condition just before or at the onset of her first few eggs. Added to that there is the problem of the cockerel's feet and claws ruining the feathers on her neck and back during mating. But these are details which can be overcome with adequate pens and housing.

I mentioned planning for exhibiting. Most top breeders think about a year ahead, and plan to have their birds in tip top condition ready for a particular show or the beginning of the show season. They will not be breeding one bird but dozens, and there will be a programme of selection starting with the eggs and continuing with incubation, chicks, growers and young adults.

Many birds will be culled or sold off. In America, some exhibitors, or stringers as they are called, buy the better birds just for showing, and go with them from show to show.
So you can see from this that there can be more to the business of showing than meets the eye, but it need not be so detailed and meticulous for the beginner.
Let us assume that you have some good birds, (you think!) that you want to show, and you have been to several shows and talked to exhibitors there. What is the next step?
In order to show you need to join a Club. There are two umbrella organisations, the National Poultry Club and the Federation of Poultry Clubs. Loosely, the National Poultry Club is more of a south of England organisation, whereas the Federation covers the north. Many people belong to both. There are numerous regional clubs and breed clubs within the two organisations so you may find yourself becoming a member of several before you embark upon showing your birds. Subscriptions should be very reasonable.
There are various sources of information on shows. Magazines such as Fancy Fowl, Smallholder and Country Smallholding all list them, and some local agricultural shows have poultry exhibitions.
Apply for an entry form from the secretary of your club and return it in plenty of time to book the number of pens you require in the relevant classes. These are normally Juvenile Hen and Juvenile Cock, ( these birds are under a year old ) Hen and Cock, ( these birds can be any age ) and Trios where a cock and two hens are exhibited together. You will then come up against the terminology Hard Feather and Soft Feather, because very often the show will be split in two with a Championship Cup going to each section. Hard Feather covers all Game birds and Soft Feather is basically everything else, although there are some exceptions! Sebrights have a hard feather but are classified as Soft, and all rare breeds are classified as Rare Breeds irrespective of hard or soft feathers!! Confusing? It certainly is to start with, but you will get the hang of it eventually!
Entry forms normally need to be in about a month beforehand, but some shows such as the National have very early closing dates. Local shows tend to be more easy going as they like to have as many entries as possible. You will need entrance passes and car park stickers, particularly for agricultural shows, otherwise you will not get near the showground.
It's a good idea to have two "teams" of birds if you can. When you have chosen them you will have to train them for exhibition work. The reason for this is to ensure that when they are being judged they are not frightened of

*Home made show cages for training show birds*

being handled or having a judging stick pointed at them. ( This is like a kind of adjustable metal "car aerial" that the judges use to move the birds round the cages and get them to stand well and show off their best points.) Remember, you want your birds to stand up in the pen and show themselves, not flutter about and peck the judge, or worse still, stand with their heads in the corner and their backs to everybody! The length of time they will have to spend in the training pen will vary according to the breed and how they have been reared; it could be as little as a couple of weeks or as long as several months. So start off by petting your birds and getting them used to being handled. Give them extra tit bits to eat, treats such as bits of lettuce or apple, peanuts or currants. People used to give their birds meat before it was banned. Get your birds used to being put headfirst into their cages and being moved round with a judging stick. Calm them down when they are in the confined space so that they get used to it and don't panic when they arrive at the show.

Now is the time to dust your birds against lice and fleas, and you can try introducing some special foods to enhance their appearance: linseed helps to give the feathers more sheen, and maize helps yellow legs to look more yellow, but be careful not to feed it to white breeds as their feathers will go a creamy colour. Buff and white breeds should be kept out of direct sunlight as the buff colour fades, and the white birds will turn a brassy sort of gold, particularly on the hackle.

The cages you need for training can be bought at the larger shows or you can make your own using wooden boxes. The sizes should be 20 x 20 x 24 inches, (50 x 50 x 60 cms) high for large fowl, and 14 x 14 x 16 inches, (36 x 36 x 40 cms) high for bantams. These measurements are not exact, but the cages must be large enough for the birds to stand up and turn round in comfortably. They also need a door on the front for access, not only for feeding but also for when you are handling your birds. If you buy the show cages, which normally come in threes, make a plywood tray with 3 inch sides round all the cages to help keep the shavings in; you may need to make a cut-out in front of the doors so they can still open properly. Each bird will require a feeding and drinking cup which are available cheaply at the larger shows. Hook these onto the wire inside the cage. There are two reasons for this: first the bird gets used to hands coming into the cage, and second, if the cups are on the outside, the bird's hackle or neck feathers might suffer by chafing on the wire.

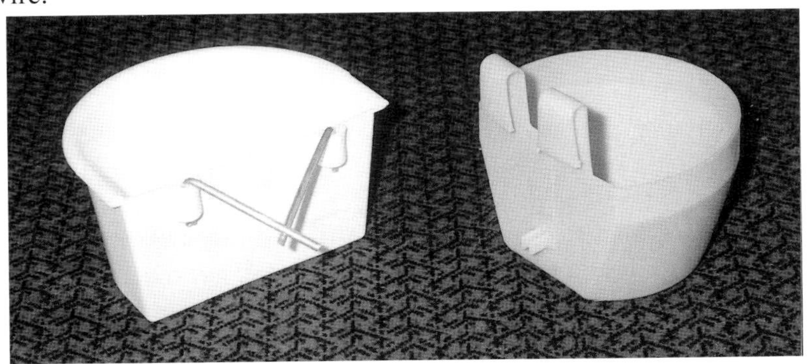

*Two different kinds of cage drinking cups*

So, the entries are in, you have trained your birds, and the show date is looming fast. Twentyfour to fortyeight hours before the day you must wash your birds. Experienced exhibitors often have their own secret lotions, potions and shampoos to help their birds look their best, and you may get some good tips from talking to other breeders. Baby shampoo works well in soft water, and there are certain dog shampoos that help to whiten coats, which are used on white birds. I have found showering a bird in the bath works best, but some people use large paint tubs. Showering is more like rain and must be less alarming than total immersion in a container.

Use warm water and try not to fluster the bird too much, as some have been known to die of heart attacks during the washing process. At the end pour on a little vinegar then rinse it off to give the bird's feathers a really good shine. ( It works well on human heads too!) Detailed care should be taken over the legs, feet and toes. Scrub them gently to remove any dirt while they are soft from the warm water, and use a cocktail stick for any stubborn bits.

Penned birds' toe nails grow because they are not being used, so you will have to trim and shape them. Use strong nail clippers and a nail file or fine sandpaper.

Towel the birds carefully and gently, trying to make this unnatural experience as pleasant as possible for them. Next use a hand held hair dryer to get rid of the worst of the dampness. The birds need to dry out slowly so that their feathers do not curl the wrong way, so put them in clean pens with fresh shavings and keep them at a warm room temperature. They can take from 12 to 15 hours to dry out completely, and during this time they will sort out their own feathers with much preening and titivating. When they are completely dry, use a silk handkerchief to give hard feathered birds a good polish. Finally, oil their legs and feet with a little olive oil.

Pull out any broken or incorrectly coloured feathers that your birds might have, but be careful not to pull out too many from the same place as the judge will certainly notice an uneven patch. It goes without saying that all the wing feathers must be perfect.

You are bound to hear some talk about faking which can involve ploys such as trimming combs and wattles and dying feathers. A classic case concerned a black hen with a patch of white feathers which the breeder had inked over using a black felt tip pen. All was revealed when the judge, dressed in his usual white coat, came to handle the bird during judging, and got marked with black ink! To pull out the odd feather is OK, but otherwise, dying or painting is just deception and cheating.

The next thing to consider is travelling boxes. You must make sure that these will all fit inside the vehicle you are going to use, and that they will be comfortable and roomy enough for your birds. Strong cardboard boxes that won't crush should do well for the novice to start off with. Don't forget to cut out some ventilation and hand holes, and also put in a good layer of clean shavings. Many breeders use plywood boxes with handles or straps. ( See "Poultry House Construction" from the Gold Cockerel Series ). The original travelling boxes were made of willow with

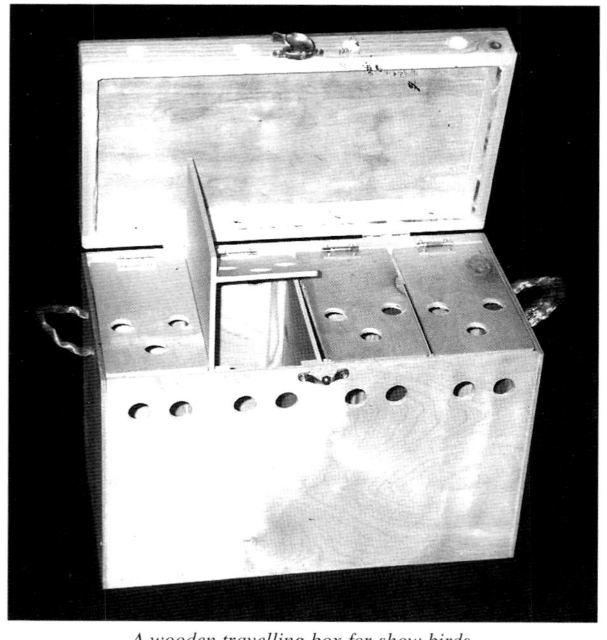

*A wooden travelling box for show birds*

sacking linings. The main idea is that your birds should arrive cool and as clean as possible. Do not put food or water in the travelling boxes. When the show day arrives, allow plenty of time to give your birds a snack before leaving, box them up and get changed yourself. You may also want to take something to eat and a thermos of tea or coffee, as, although most shows have catering facilities, they can be expensive and some shows don't provide anything.

The question of food is all important for the birds as well. They don't want bulging crops which will ruin their shape, but on the other hand they shouldn't be so weak with hunger that they can't stand properly. There are many light snacks that you can give them which will keep them going until after the judging is completed, normally by lunch time. These snacks can be small quantities of broken biscuits, crushed peanuts, currants or sprouted oats. Don't rush to feed your birds after the initial judging, because if one of them has come first in its class it might be put up for champion in its breed and / or champion in its type, Hard Feather or Soft Feather; so make sure all the judging is over before giving your birds a good square meal. Whole wheat or pellets are normally provided by the show staff, and there should be a water drinker in the cages at all times. Some people add vitamins to the water because during showing, birds loose weight and condition due to stress, particularly at the longer shows.

When you arrive make sure you get safely parked. Some small events take place in Village Halls or School Halls where there is not much parking, and of course traffic wardens don't differentiate between shoppers and exhibitors! This is where leaving plenty of time is crucial to the operation so you can concentrate on your birds with no worries about your vehicle.

*Show cages in Holland. The judging card above lists the good and poor points of the bird*

When you are in, check your entry form and find your pens; the numbers will be on labels attached to them. (I always used to write the pen number on the box holding the relevant bird to make things easier.) You may need to find sawdust to put in the pens, but this will depend on how well the show is organised. Put the finishing show touches to your birds as you pen them. This will include re-aligning any feathers, oiling legs and feet with olive oil and putting Vaseline on the combs, faces and wattles to give them all more allure.

Drinking cups are normally provided but if not, put in your own with a symbol or mark or initials on the bottom so they can't be seen by the judge or steward. If they can be seen the bird will be eliminated. The mark or initial should ensure that your cups don't get pinched.

Show days can be rather long, especially if you got up at the crack of dawn. They can also be rather boring, so take something to read or go and visit a nearby attraction once you have settled your birds in their cages. ( After lunch when the judging is over, things do start to live up though. ) Alternatively you could offer your services as a steward to one of the judges and learn from him. There are never enough people to help out at a show, particularly at the end when all the cages and tables have to be cleared away. Whatever you decide to do, it is always exciting to come back and see how your birds have fared.

A word or two about judging. Proper judges do take exams so that they can be tested on their abilities, but they are only human and do sometimes make mistakes which they have to live with. You will find that some of them have

*This is a view of a Dutch Poultry show where the birds are beautifully staged*

preferences for certain breeds and certain types within those breeds. Experienced exhibitors will only show certain birds to be judged by certain judges. If your bird has not done well it is always worth asking the judge why after the judging is over, and if he is worth his salt he will be fair and frank with you. Judging rare breeds can be very difficult and requires experience and knowledge. The judges are usually listed in the entry forms.

I think it must be stressed here that showing frequently seems to bring out the worst in people today, and there are many who change when they find themselves in a competitive situation, particularly if their bird is passed over for the top prize. Showing should be fun, and in the old days most exhibitors showed not only their poultry, but their vegetables and flowers as well. There used to be a feeling of friendly competition because exhibitors knew that they might not win that year or under such and such a judge, but there would always be another chance another year. Many breeders used to encourage the new and the young into the hobby, but sadly today there seems to be a rather more selfish attitude.

It is a sad reflection on our times that these days so many of the cages at shows have padlocks on them. Also, many of the top exhibitors have to tattoo their birds. If you do wish to lock up your birds, do it after the judging as they could be disqualified if they are locked in.

Having read all this, particularly the last few paragraphs, you may wonder why people show at all. Here are several excellent reasons for doing so:

A It is good to see your birds being judged against the best in your area.

B It is a useful way to advertise your stock. After the judging you can put up a card naming the breed, (shows are not very good at doing this), also your name and telephone number.

C You can meet some extraordinarily nice people who share your interests at shows.

D It is gratifying to receive awards for all the hard work and planning you have put in over the past year.

# NOTES